HC

Northern Lights

A practical travel guide

Polly Evans

www.bradtguides.com

Bradt Travel Guides Ltd, UK
The Globe Pequot Press Inc, USA

AUTHOR

Polly Evans is an award-winning journalist and writer. She's the author of Bradt's *Yukon*, as well as five narrative travel books, the most recent of which, *Mad Dogs and an Englishwoman*, tells the story of her learning to drive sled dogs in northern Canada. When not on the road, Polly lives in Berkshire.

Third edition published September 2017
First published September 2010
Bradt Travel Guides Ltd
IDC House, The Vale, Chalfont St Peter,
Bucks SL9 9RZ, England
www.bradtguides.com
Print edition published in the USA by
The Globe Pequot Press Inc,
PO Box 480, Guilford, Connecticut 06437-0480

Text copyright © 2017 Bradt Travel Guides Ltd
Maps copyright © 2017 Bradt Travel Guides Ltd
Photographs copyright © 2017 Individual photographers (see below)
Specialist contribution: Sheridan Williams
Project Manager: Laura Pidgley

ISBN: 978 1 78477 071 6 (print)
e-ISBN: 978 1 78477 528 5 (e-pub)
e-ISBN: 978 1 78477 429 5 (mobi)

British Library Cataloguing in Publication Data
A catalogue record for this book is available from the British Library

Photographs
Alamy: Jim Keir (JK/A), NASA Archive (NA/A); Canadian Tourism Commission (CTC); Discover the World: Hotel Ránga (HR/DTW), Kristen Folsland Olsen (KFO/DTW), Marcela Cardena (MC/DTW), Ragnar TH Sigurdsson/arcticimages.com (RTHS/DTW); NASA (NASA); Hurtigruten (H); www.kirunalapland.se (KL); Mick Mackey (MM); Mývatn Nature Baths (MNB); North Norway Tourist Board: Lunde Ingvaldsen (LI/NNTB); Konrad Konieczny (KK/NNTB), Baard Loeken (BL/NNTB), SuperStock (SS); www.travelyukon.com (TY); Visit Finland: Konsta Punkka (KP/VF), Tiina Törmänen (TT/VF); Visit Greenland: Mads Pihl (MP/VG), Paul Zizka Photography (PZP/VG); Visit Rovaniemi/Rovaniemi Tourism & Marketing Ltd (VR); Visit Sweden: Asaf Kliger/www.nutti.se/imagebank.sweden.se (AK/VS); Wikimedia Commons: Andrew Shiva (AS/WC)

Front cover Bradt edition: the northern lights over Tromsø, Norway (KK/NNTB)
Title page Reindeer sledding (AK/VS); Northern lights over Ylläs, Finland (KP/VF); Snowshoeing at Illulissat, Greenland (MP/VG)
Back cover Northern lights over the ICEHOTEL, Sweden (AK/VS); dog sledding in Canada (CTC)

Maps David McCutcheon FBCart.S

Typeset by Ian Spick, Bradt Travel Guides Ltd
Production managed by Jellyfish Print Solutions; printed in Turkey
Digital conversion by www.dataworks.co.in

Contents

FEEDBACK REQUEST AND UPDATES WEBSITE

At Bradt Travel Guides we're aware that guidebooks start to go out of date on the day they're published – and that you are out there in the field doing research of your own. So why not send us your updates? Contact us on ☎ 01753 893444 or e info@bradtguides.com. We will forward emails to the author who may post updates at www.bradtupdates.com/northernlights. Alternatively you can add a review of the book to www.bradtguides.com or Amazon.

LIST OF MAPS

Introduction

One night a few years ago I went on an evening visit to a commercial northern lights viewing centre in Yellowknife, in Canada's Northwest Territories. I didn't much want to go. I'd spent several months living in the far north, researching a book about dog sledding, and I'd seen the northern lights plenty of times. I'd become a bit blasé about tourist outings given that I'd watched the aurora regularly from the comfort of my own cabin. That evening, I preferred the idea of curling up beneath the duvet and going to sleep. But I was working as a journalist, specialising in the Arctic and sub-Arctic, and I had to.

And so began one of the most remarkable evenings of my life. That night, the northern lights put on the most extraordinary display, the likes of which I had never imagined. First they crept up like green bony fingers from the horizon, as I'd seen many times before. But then they grew and swirled until they swept across the entire sky. They seemed to take on every possible form. At times they unfurled from the horizon like a flower at dawn. Then they'd stretch out into an arched streamer, and undulate like a flag flapping in the breeze. At times they seemed to climb up into a three-dimensional tepee which collapsed downwards like a pile of luminous pick-up-sticks. The aurora was pale green with a scarlet underbelly, and it danced across the starry black sky with a speed I'd never thought possible.

I'd climbed to the top of a small hill, set up my tripod, and begun to take photographs. The pictures were beyond my photographic dreams – but I could only snap one segment of sky, then another, and I couldn't record the wild energy of the aurora as it tumbled across the sky. And so I stopped pressing the button. Instead I lay down on my back in the snow, and allowed myself to be consumed by the show overhead. It lasted just 10 or 15 minutes but provided perhaps the most intense experience of beauty that I have ever known.

Many people travelling in quest of the northern lights won't catch a display of this level. Some won't manage to make their visit coincide with any auroral activity at all. But even if you're unlucky with the lights, a trip to the Arctic in winter will provide memories that, in my opinion, outstrip any beach holiday. There's something about the blue light of the mornings that turns buttercup yellow at noon, then takes on pinky tinges as evening falls, the serenity of the pure white landscapes, the impossibly intricate design of a single snowflake, the sparkling jewels of hoar frost that grow on the branches of spruce, that never ceases to entrance. In the daytime, there's dog sledding, skiing and snowshoeing for the energetic, and snowmobiling for adrenalin enthusiasts; as for the night, the far north is home to an ever-increasing array of ice hotels, ultra-chic igloos, and cosy log cabins with roaring wood stoves. I hope this booklet will help you to learn a little not just about the northern lights, but also about the wider experience the northern countries offer. And I hope that you, like me, will find your Arctic journey so mesmerising that, across the years, you can't help but return.

What Are the Northern Lights?

The ancient Greenlanders thought the northern lights were a sign from the heavens that their ancestors were trying to contact the living. The Norwegians saw them as old maids dancing. Modern science is rather more prosaic. It tells us that the northern lights are created by charged solar particles ejected from the Sun in a solar wind during explosions or flares. If these particles reach the Earth – a journey that takes two or three days – they are diverted around the planet by the Earth's magnetic fields. (If they were allowed to penetrate the Earth's atmosphere, they would damage us with their intense radiation.)

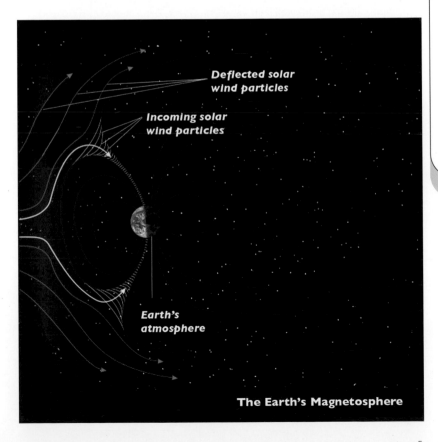

Deflected solar
wind particles

Incoming solar
wind particles

Earth's
atmosphere

The Earth's Magnetosphere

Literally, aurora borealis translates from the Latin as 'northern dawn'. There's some dispute as to who first coined the expression. Some attribute it to Italian scientist Galileo Galilei who saw the major aurora of 12 September 1621 (page 45); others claim French astronomer Pierre Gassendi first used the phrase following the same event. Either way, the reference to dawn alludes to the red colour that is seen in the aurora at the lower latitudes at which both men would have observed this impressive display.

The southern lights (page 22) are sometimes referred to by their Latin name 'aurora australis'. Together, the northern and southern lights can be called the 'aurora polaris'.

Most of the solar wind travels around the Earth and disappears into space. Some particles, however, enter the Earth's upper atmosphere at its polar regions. As they do so, they collide with atoms and molecules there, which absorb some of the solar particles' energy. This is referred to as 'exciting' the atom or molecule. In order to return to their normal state, the atoms and molecules emit photons, or light particles. The northern lights are therefore an actual source of light, not a reflection of sunlight as some people suppose.

The colour of the northern lights depends on which gas the solar particles collide with. In the auroral displays that are most commonly to be seen with the naked eye, oxygen emits a greenish-yellow burst of light, while the photons produced by nitrogen are crimson. This activity takes place in the Earth's high atmosphere, at 100km or more above the planet's surface.

The aurora borealis as seen from space (NA/A) page 8

The aurora appears red at higher altitudes, and green at over 100km (LI/NNTB) see below

THE COLOURS OF THE AURORA

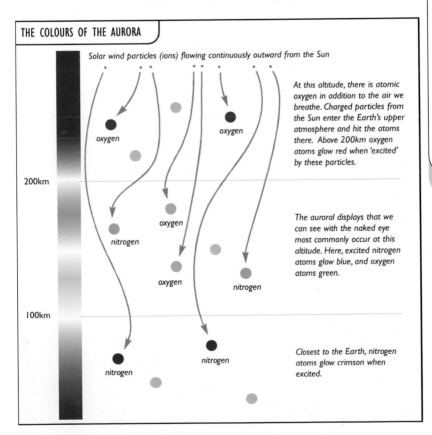

Solar wind particles (ions) flowing continuously outward from the Sun

At this altitude, there is atomic oxygen in addition to the air we breathe. Charged particles from the Sun enter the Earth's upper atmosphere and hit the atoms there. Above 200km oxygen atoms glow red when 'excited' by these particles.

The auroral displays that we can see with the naked eye most commonly occur at this altitude. Here, excited nitrogen atoms glow blue, and oxygen atoms green.

Closest to the Earth, nitrogen atoms glow crimson when excited.

200km

100km

oxygen

oxygen

nitrogen

oxygen

oxygen

nitrogen

nitrogen

nitrogen

MAGNETIC VERSUS GEOGRAPHIC POLES

The Earth's magnetic field lines, which help to create the northern lights by diverting solar particles towards the poles, run between its magnetic rather than geographic poles. So, while the geographic North Pole sits permanently at 90° latitude, and is the point at which all the lines of longitude converge, the Earth's magnetic North Pole moves: it is influenced by the perpetually shifting molten iron that makes up the Earth's outer core. At the time of writing the magnetic North Pole is in the Canadian Arctic, and is calculated to be drifting towards Russia at a speed of about 60km per year.

THE AURORAL OVALS

The convergence of the Earth's magnetic fields around the magnetic North and South poles lead the northern lights to exhibit in auroral ovals, or rings around the top and bottom of the Earth. If you want to see the northern lights, you therefore have to be beneath, or within sight of, the auroral oval.

The auroral ovals stay in a fairly fixed location in space while the Earth rotates beneath them. When levels of solar activity are low, the northern auroral oval sits between 60° and 70° of latitude; it is wider on the night-time side of the Earth. During a very violent solar flare, however, the auroral oval fattens or bulges and the aurora can be seen from lower latitudes. However fat or thin, the auroral oval is always present but can only be seen in its entirety – as a ring – from space.

The auroral oval encircles the magnetic North Pole (NASA)

THE SUNSPOT CYCLE

Sunspots are dark blotches on the Sun's surface that are cooler than the surrounding regions (though they're not exactly cool – if a sunspot could be suspended in the night sky it would still provide more light than the full moon). Sunspots are temporary phenomena, which contain concentrated magnetic field lines; this intense magnetic activity prevents them heating to the same temperature as the rest of the surface of the Sun. Solar disturbances are greater when sunspot numbers are high – and so, in turn, we see more intense and active northern lights when sunspots are common.

Sunspots come and go in 'sunspot cycles' which peak about every 11 years. During these peaks, solar flares are more common and more energetic, and so the northern lights – though visible at all stages of the sunspot cycle – tend to be more frequent and intense at these times. Additionally, the years following a peak see increased northern lights activity. This is because coronal holes (areas of the Sun that are low in density and have open magnetic fields – the magnetic fields lead out into space rather than looping back into the Sun, and so allow the solar wind to escape from the Sun) form during sunspot cycle peaks, and they too lead to greater auroral activity.

THE DIFFERENT SHAPES OF THE NORTHERN LIGHTS

The aurora can be seen as undulating ribbons and shimmering curtains, spiky fingers that creep up from the horizon and dazzling rods of colour that burst like a firework from a single point high in the sky. The different shapes are partly caused by the position of the aurora in relation to the magnetic zenith – the point in the sky that you can see if you look along the line of the Earth's magnetic field. If the aurora runs across the observer's magnetic zenith, the observer will see waves or lines converging into a single spot. If the aurora is some distance away from the observer's magnetic zenith, it will look more like a two-dimensional curtain or line. It works a bit like perspective when you stand at the foot of a tall building: looking directly upwards, you see the nearest building's walls converge more sharply than those of buildings some distance away.

CAN YOU HEAR THE NORTHERN LIGHTS?

Indigenous northern people and observers from the lower latitudes have frequently told of hissing, crackling or rustling noises that accompany the aurora. Until recently, scientists couldn't explain this: the northern lights are located in the Earth's upper atmosphere, where the air is too thin to carry sound waves. However, a recent study by scientists at Finland's Aalto University reckons that the solar particles – or the geomagnetic disturbance created by them – generate sound when they're much closer to ground. Don't expect to hear them during your aurora-watching though. The soundtrack doesn't play alongside every aurora display, and when it does, it is brief and faint. Your best bet for hearing the northern lights is from the comfort of your very own sofa – with an internet search you'll find various websites that claim to have recorded them.

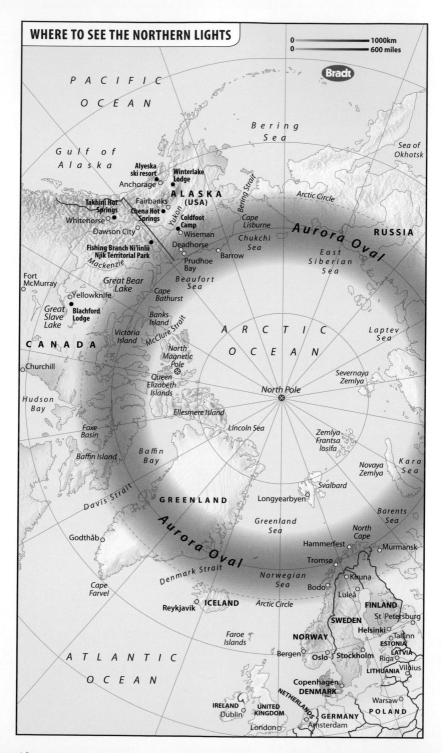

WHERE TO SEE THE NORTHERN LIGHTS

0 _____ 1000km
0 _____ 600 miles

Bradt

PACIFIC OCEAN

Gulf of Alaska

Bering Sea

Sea of Okhotsk

Alyeska ski resort
Winterlake Lodge
Anchorage
ALASKA (USA)
Fairbanks
Takhini Hot Springs
Chena Hot Springs
Whitehorse
Dawson City
Yukon
Coldfoot Camp
Wiseman
Deadhorse
Prudhoe Bay
Barrow

Bering Strait
Arctic Circle
Cape Lisburne
Chukchi Sea
Aurora Oval
RUSSIA
East Siberian Sea

Fishing Branch Ni'iinlii Njik Territorial Park
Mackenzie

Fort McMurray
Yellowknife
Great Bear Lake
Cape Bathurst
Beaufort Sea
Laptev Sea

Great Slave Lake
Blachford Lodge
Banks Island
Victoria Island
McClure Strait
North Magnetic Pole
CANADA
ARCTIC OCEAN
Severnaya Zemlya

Churchill
Queen Elizabeth Islands
North Pole

Hudson Bay
Ellesmere Island
Lincoln Sea
Zemlya Frantsa Iosifa
Novaya Zemlya
Kara Sea

Foxe Basin
Baffin Island
Baffin Bay
Svalbard
Longyearbyen
Barents Sea

Davis Strait
GREENLAND
Greenland Sea
North Cape
Hammerfest
Murmansk

Godthåb
Aurora Oval
Tromsø
Kiruna

Cape Farvel
Denmark Strait
Norwegian Sea
Bodø
Luleå
FINLAND
St Petersburg

Reykjavík
ICELAND
Arctic Circle
SWEDEN
Helsinki
Tallinn
ESTONIA
LATVIA
Riga
LITHUANIA
Vilnius

Faroe Islands
NORWAY
Bergen
Oslo
Stockholm

ATLANTIC OCEAN

Copenhagen
DENMARK
NETHERLANDS
Warsaw
POLAND

IRELAND
Dublin
UNITED KINGDOM
GERMANY
Amsterdam
London

2

Where to See the Northern Lights

It doesn't follow that the further north you travel, the more likely you'll be to see the northern lights. This is because the aurora is most visible beneath the ring known as the auroral oval (page 8), which, though constantly shifting, usually encircles the Earth between 60° and 70° of latitude.

SWEDEN

Northern Sweden is characterised by its tremendous expanses of pine forests and mountains. Much of this land is protected in national parks, and makes an almost limitless playground for outdoors aficionados in both summer and winter – Abisko and its surroundings are particularly recommended (page 13). Accommodation throughout Swedish Lapland, as in the rest of Scandinavia, is bright, clean and reliable. Many visitors head for at least one night to the glittering ICEHOTEL (page 39), which has helped to put the area around Kiruna firmly on the tourist map.

THE NORTHERN LIGHTS IN SWEDEN The best hub to head for in Sweden is Kiruna, which is now linked to London with Discover the World's direct flights. From there

BELOW LEFT: Sweden's original ICEHOTEL is still a major draw for visitors seeking spectacular aurora displays and intricate ice sculptures (RTHS/DTW) page 39

BELOW RIGHT: Kiruna's church will be moved piece by piece as the town is relocated (KL) page 33

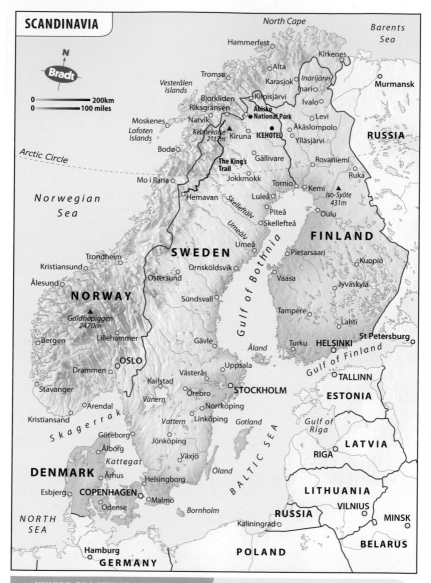

WHERE EXACTLY IS LAPLAND?

The term 'Lapland' is sometimes used to refer to the Swedish province of Lappland, or the now-defunct Finnish province of Lappi. Usually, however, Lapland is considered to span the northern parts of four countries: Norway, Sweden, Finland and the westernmost sliver of Russia's Kola Peninsula. It refers to the land traditionally inhabited by the Sámi people (page 34), who used to be called the Lapps (originally a pejorative term, this word is now little-used); the Sámi are now a minority even in their far northern homeland. The Sámi people call their traditional lands Sápmi, and you'll often see this name used instead of Lapland.

it's an easy journey by road to the dazzling sculptures of the ICEHOTEL (page 39), and to **Abisko** (below), probably the best location in Europe for seeing the northern lights. Abisko is also on the King's Trail (page 31), with its fantastic cross-country skiing opportunities.

Abisko Mountain Station Abisko should be right at the top of the list of UK-based northern lights seekers for two good reasons. Firstly, it's simple to get to: Abisko is 100km west of Kiruna and easily accessible by road, and Kiruna can be easily reached by direct flights from London. Secondly, Abisko's geographical location means it enjoys a higher-than-average number of cloudless nights – thanks to the surrounding mountains that deter the rain clouds, Abisko is the driest place in Sweden.

Abisko makes the most of its clear nights with the Aurora Sky Station, a dedicated northern lights viewing facility that sits on top of a mountain giving huge views of the sky. The northern lights can be seen on about 50–60% of the nights that the Sky Station is open, from December to March. A chair-lift takes you up to the summit where there is a café and northern lights exhibition. It is also possible to enjoy a special four-course meal here. Abisko Mountain Station, at the bottom, has simple but comfortable rooms. You can combine a stay here with the nearby ICEHOTEL (page 39), a 1½-hour transfer away.

NORWAY

Norway is known for its fjords and perfectly picturesque fishing villages such as those of the **Lofoten Islands**. The Lofotens, whose red-painted houses huddle beneath soaring granite cliffs, are fabulous in winter as well as summer and temperatures are often warmer here than in other parts of Lapland. If you're looking for an authentic touch to your stay, try sleeping in a *rorbu* – a fisherman's hut on stilts. Many are now available for rent to tourists; try to find a rickety old one rather than a shiny new imitation.

NORWAY AT A GLANCE

Language Norwegian (Bokmål and Nynorsk)
Population 5.1 million
Currency Norwegian krone (NOK)
Time GMT +1 (winter), GMT +2 (summer)
Electricity 220V; plugs have two round pins

Aurora-seekers in Norway can view the lights from Hurtigruten's ships (H)

The aurora regularly lights up Norway's peaks and fjords during the winter months (LI/NNTB)

WHERE IS FINNMARK?

Confusingly, Finnmark has nothing to do with Finland, but is the northeasternmost province of Norway. Norway's North Cape sits at Finnmark's tip.

The main city of northern Norway is **Tromsø**; surrounded by mountain peaks on one side and the ocean on the other, it blends historic timber houses with a lively contemporary restaurant scene.

Stormily scenic, Europe's most northerly point is Norway's **North Cape (Nordkapp)** at 71°N. You can't drive your own vehicle up here in winter as the road is closed; you have to take the coach tour that follows in the wake of a snow plough. The North Cape's polar night lasts from mid-November to the end of January.

THE NORTHERN LIGHTS IN NORWAY The most scenic way to search out the northern lights is by sailing along Norway's coastline. Hurtigruten (page 13) offers voyages of various lengths and its ships are furnished with observation lounges. Alternatively, you can fly easily to Tromsø, which enjoys an outstanding location for northern lights viewing as it's right beneath the auroral oval and, if you wish, take a short coastal day or half-day whale-watching trip from there. Alternatively, snowmobiling and dog sledding are great activities to enjoy by moonlight while keeping your fingers crossed for a light show.

Svalbard, an archipelago perched high above Scandinavia at about 80°N, is part of Norway. Its main island is Spitsbergen; the largest settlement is Longyearbyen, which has a human population of around 2,200 – and about the same number of polar bears. A winter journey to Svalbard constitutes a genuine adventure. It's certainly not the easiest destination from which to see the northern lights – to start with, the sun doesn't rise between November and mid-February. While this could be a bonus – the extended hours of darkness increase your chances of seeing the lights, after all – most winter travellers prefer to visit Svalbard from February onwards. At this time of year they still stand a good chance of seeing the aurora yet can enjoy, for example, a multi-day snowmobiling expedition with a little bit of daylight thrown in.

FINLAND

Finland is the most sparsely populated country in the European Union and around a million, or about 20%, of its population lives in the cosmopolitan and cultural capital, Helsinki. The countryside is characterised by its thousands of lakes and islands, and landscapes covered in lush pine, spruce and birch forests. The largest city in Finnish Lapland is **Rovaniemi**, hugely popular with winter tourists for its Santa Claus attractions (page 35), as well as its many museums, restaurants and cafés.

In winter, Finland attracts snow-sports enthusiasts: the country has around 80 ski resorts and tens of thousands of kilometres of marked trails for cross-country skiers. In the north of Finland you can ski from November to May. **Ylläs**, the largest ski resort in Finland, is a haven for downhill skiers and snowboarders. For downhill, also try **Iso-Syöte** in southern Lapland, where authentic log cabins near the slopes provide holiday accommodation with a good chance of northern lights viewing.

THE NORTHERN LIGHTS IN FINLAND There's more to Finnish Lapland than Santa Claus. Top spots for aurora aficionados include **Lake Inari** on the Russian border where **Nellim Wilderness Hotel** sits on the lakeshore amid the taiga forest. Its 'Aurora Bubble' and 'Aurora Kota' are specially designed for viewing the lights – they have windows facing the north sky, and beds for when the lights have gone to sleep. Sister hotel **Muotka Wilderness Lodge** is located in the taiga forests of Finland's Saariselkä fells and Urho Kekkonen National Park. In the daytime, guests can take to the ski trails or dare each other to dip in the pool cut from the ice of the nearby river.

If ice swimming doesn't burn enough calories, try the daytime activities at **Iso-Syöte Winter Hideaway**. Here you can clip on your downhill skis – Iso-Syöte is home to some of Finland's best slopes. Cross-country skiing, showshoeing, husky sledding, snowmobiling, reindeer safaris and ice fishing are also on offer as well as trips to Rovaniemi's Santa Claus Village and Ranua Wildlife Park.

Iso-Syöte is a favourite ski area for the Finns (TT/VF)

The mighty Seljalandsfoss is one of the most-visited waterfalls in Iceland (RTHS/DTW)

ICELAND

Iceland may not have many people – it is the most sparsely populated country in Europe – but its volcanoes have a habit of drawing attention to themselves. In 2010, ash from Eyjafjallajökull caused flight disruption across Europe, while the eruption of Grímsvötn the following year sent a cloud of ash and lava 20km into the atmosphere. It's certainly not the case, however, that Iceland is hunkering miserably beneath a cloud of volcanic ash. This country moulded by geothermal activity is a place of pristine beauty where geysers, volcanic peaks and lava fields create spectacular sights. Sightseeing flights take visitors over the volcanoes, weather conditions permitting, while the more energetic can take a scenically stunning glacier hike. You should always take a guided tour or hire a private guide for such walks. Indoors, boutique hotels and culinary gurus combine to create a high-end escape for those who like life comfortable. Direct flights from London to Keflavík (Iceland's main international airport) take about 3 hours, making Iceland a highly accessible yet serenely exotic destination.

> **ICELAND AT A GLANCE**
>
> **Language** Icelandic
> **Population** 330,000
> **Currency** Icelandic króna (ISK)
> **Time** GMT year-round
> **Electricity** 220V; plugs have two round pins

THE NORTHERN LIGHTS IN ICELAND Yes, you can see the northern lights in Reykjavík if the display is intense, but the light pollution of the city prevents you from seeing the show at its best. And you don't have to travel far to escape artificial light. Hotel Rangá (pages 39–40), for example, is located around a 2-hour drive from the city.

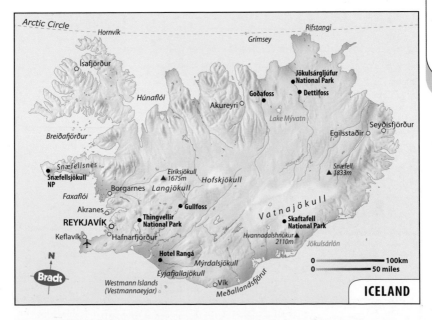

GREENLAND

How can I put this nicely? Greenland's capital, Nuuk, enjoys *average* annual temperatures of −1.6°C. And Nuuk is in the south of the country. Pack a good woolly hat, however, and Greenland is a stunning destination both for its mind-boggling beauty and its downright differentness. To start with, 85% of the country is covered in ice – Greenland's icecap covers an area more than three times the size of France. You can access one little corner of it very easily from Kangerlussuaq, home to Greenland's international airport – the edge of the icecap is just 25km away on a good road. Old Arctic hands can ski or snowmobile across its width. The rest of us can take an easy stroll across a tiny section (the inexperienced should go with a guide), then try out Greenland's husky sledding and musk oxen safaris. Travel up to Ilulissat (home to 4,000 people and 5,000 sled dogs) and you'll see the electric-blue icebergs for which Greenland is famous. Ilulissat's icefjord is one of the few places where the icecap meets the sea, and its glacier is one of the most actively calving in the world.

THE NORTHERN LIGHTS IN GREENLAND Greenlandic legend relates that, when the northern lights are dancing in the sky, it means the dead are playing football with a walrus skull. Take it or leave it, Greenland's northerly location makes it an ideal spot for aurora sightings and, as its population is so sparse, it's easy to get away from the city lights. A word of warning: parts of Greenland are home to healthy populations of polar bears – don't wander off into the countryside to escape light pollution without taking local advice first.

RIGHT: Greenland's location means good chances of an aurora sighting, even in the towns (MP/VG)

BELOW: Take a boat to Ilulissat to see Greenland's famous icebergs (PZP/VG)

ABOVE LEFT: A tundra buggy in Churchill allows visitors to meet polar bears safely (CTC) page 38

ABOVE RIGHT: Thermal waters keep winter visitors warm at Chena Hot Springs in Alaska (DF/CHS) page 37

CANADA

Northern Canada is home to some of the world's last wildernesses, where caribou, bears and wolves outnumber human residents. In winter, temperatures dip low, but many insist that this is their favourite time of year: there's an intense beauty to the white silence that reigns across the Canadian north in the snowy months. The auroral oval cuts across the whole of northern Canada, making it a top destination for northern lights tourism; this guide focuses on the western side of the country. During daylight hours, there's dog sledding, snowmobiling and snowshoeing for outdoors enthusiasts, while Churchill is the world's top destination for polar bear enthusiasts (page 38).

> ### CANADA AT A GLANCE
>
> **Language** English and French
> **Population** 35,160,000
> **Currency** Canadian dollar (CAN$ or CAD)
> **Time** Time zones vary between GMT –3.5 and GMT–9
> **Electricity** 110V; plugs have two flat pins or two flat pins and a round grounding pin

THE NORTHERN LIGHTS IN CANADA **Yellowknife** is renowned for its northern lights (it's here that I experienced the most spectacular display I've ever seen); excursions include an aurora-viewing evening from a cabin that visitors reach by snowmobiling across Great Slave Lake. For a more remote experience still, fly to the eco-lodge at Blachford Lake (page 40) where you can try your hands and feet at skating, ice fishing, dog sledding and snowmobiling. Far from artificial light, it's an ideal location for aurora gazing. Alternatively, in **Whitehorse**, you can mix your aurora-viewing with dog sledding and sightseeing by air, then lie back in the Takhini Hot Springs and watch the sky turn green while your hair grows white with icicles. In Churchill, polar bears pose for photos in the daytime (page 38). Alternatively, for a truly wild bear-viewing experience, take a helicopter to Yukon's remote **Fishing Branch Ni'iinlii Njik Territorial Park** where small groups can see the grizzly bears that congregate during the autumn to feed on the spawning salmon – despite sub-zero conditions, water flows here year-round due to thermal energy stored in underground reservoirs. And once the bears have gone to bed, look up to the skies to see if the aurora's coming out to play.

ALASKA

'The odds are good, but the goods are odd.' So say Alaskan women in search of a man. For many visitors to the USA's frozen north, this is the charm of the place. The Americans here might still have good teeth and wish you a nice day, but there's a rugged down-to-earth character to the Alaskans. Here, eccentricity is not only tolerated but celebrated, and men in hard-wearing dungarees guffaw in the face of the southern states' fashionistas. But if the

ALASKA AT A GLANCE

Language English
Population 737,000
Currency US dollar (US$)
Time GMT –9 (summer), –8 (winter)
Electricity 110V; plugs have two flat pins or two flat pins and a round grounding pin

humans can be a little funny-looking, Alaska's natural world is a dazzler, with mountain ranges, glacier fields and tundra home to moose and caribou, wolves and Arctic foxes. During the winter, daytime activities include dog sledding, snowmobiling and snowshoeing as well as all manner of skiing experiences including alpine, cross-country, back-country, heli-skiing and snowboarding.

Alaska's economy revolves around the oil and gas, fishing and tourism industries. The former invites a really fascinating tourist experience, with trips to Deadhorse, the community that serves those that work at the Prudhoe Bay oil fields, via the Dalton Highway and Coldfoot Camp.

THE NORTHERN LIGHTS IN ALASKA Of Alaska's towns, **Fairbanks** is the top spot for northern lights viewing thanks to its location beneath the auroral oval; 100km outside town, the lovely Chena Hot Springs resort offers aurora viewing either from the hot springs themselves, or from the surrounding wilderness. Having come this far, however, it's worth a little extra travelling into outback Alaska. **Coldfoot Camp** (so named because some gold-rush stampeders got 'cold feet' here, and turned around for home) lies 175km north of Fairbanks on the Dalton Highway – a road that cuts through scenery so extraordinary that it's really a destination in itself. If you're looking for boutique loveliness and a chocolate on your pillow, Coldfoot Camp isn't the spot for you: essentially, it's a truck stop serving the drivers that haul supplies up to the oilfields at Prudhoe Bay. As such, it's full of interesting characters with forests of facial hair. It's not just the people-watching that makes this a fascinating stop for the night, though. A trip to the nearby village of Wiseman (population 12) offers an insight into a world away from cellophane-wrapped joints of meat and evenly shaped vegetables, and the people's love and understanding of the land they live in left me, as an urbanite, in genuine awe. Come here in the evening, and they'll conjure up a northern-lights show – weather permitting, of course.

THE SOUTHERN LIGHTS

Auroral activity occurs around the magnetic South Pole just as it does around the north. However, because the southern auroral oval lies over frigid ocean and uninhabited ice pack, the southern lights, or aurora australis, are seen far less frequently than their northern counterparts (by humans at least – it's fair to assume the penguins regularly enjoy a good show). The southern lights are sometimes observed in the most southerly parts of Argentina, Chile, New Zealand and Australia, but you stand a better chance of catching a good display in the north.

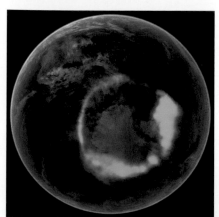

LEFT: Pictures from space show the auroral oval around the South Pole (NASA) page 8

BELOW: The aurora australis swirls over Halley Research Station in Antarctica (JK/A)

- You can only see the northern lights when the sky is dark. During the light nights of the Arctic summer, the aurora may be active – but it won't be visible because the light emitted by the aurora is much weaker than sunlight.
- You can only see the northern lights when the sky is clear of cloud. Some people claim the aurora comes out when temperatures are colder. This isn't the case – it's just that when the skies are cloudless, temperatures tend to drop.
- The northern lights are most commonly seen between 17.00 and 02.00. They don't usually exhibit for long – they may show only for a few minutes, then glide away before returning. A good display may last for no longer than a quarter or half an hour, though, if you're really lucky, it could extend to a couple of hours or longer.
- The aurora is at its most active around the equinoxes – that's to say, in March and September. The end of August and early September are good times to travel if you're a keen hiker, yet still want to try your luck with the lights – at this time of year the snow hasn't fallen, but the night-time skies are dark.
- The northern lights become more active and intense around the peak of a sunspot cycle, and in the three to four years immediately following the peak (page 9).
- The waxing and waning of the moon makes no difference to the northern lights. While a full moon lightens the sky, and may therefore reduce the visual intensity of a display, the northern lights can be seen at all stages of the moon's cycle.

POLAR NIGHTS The Arctic Circle sits at 66°33'N. At this point of latitude the Sun doesn't rise on the day of the winter solstice (and doesn't set at the summer solstice), though, as the Sun is hovering just below the horizon, the Arctic Circle never goes completely dark – there's always a bit of dusky twilight. The further north into the Arctic Circle you go, the longer the periods of winter darkness.

A group of tourists enjoys a night-time light show overhead (MM)

Finland's log cabins provide cosy accommodation for aurora-seekers (TT/VF)

The Lofoten Islands (named 'lynx foot' by the Vikings after their supposed resemblance) are known for their spectacular natural light (BL/NNTB)

3

Practical Information

TOUR OPERATORS

A number of tour operators include northern lights viewing trips among their worldwide portfolios. Go to www.bradtguides.com/nltours for a list of recommended operators.

WHAT TO WEAR

With the right clothing, the Arctic winter can be surprisingly comfortable. Many local suppliers will lend or rent you the thick outer garments that are expensive to buy. If your holiday consists of organised excursions such as northern lights viewing plus short dog-sledding and snowmobile trips, good-quality ski clothing should see you through (tip – check out summer sales in the UK for bargains on winter clothing, especially those marked 'thermal'). However, if you're going to be spending extended periods outdoors in sub-Arctic and Arctic weather, you should upgrade to a higher level of protection. If you do need to buy your own outer garments, one of the best suppliers of down clothing in the UK is Rab (*www.rab.uk.com*).

THE LAYER PRINCIPLE It is much better to wear a number of thin layers than just a few thick ones. The air trapped in between thin layers warms to your body's temperature and acts as valuable insulation. Make sure your clothes fit well and that some of your layers are of differing sizes to avoid constriction, which will prevent air circulation and will be uncomfortable.

WHAT TO BRING

Underwear In cold conditions, it's better to wear wool, silk or synthetic polypropylene next to your skin. Avoid cotton: when you sweat, it gets cold and clammy, and doesn't dry out easily. Merino wool is excellent – try the Chocolate Fish Merino, Icebreaker or Ulfrotte brands. On top of your base layer, you'll need to wear at least two or three additional layers, which should be made of fleece or wool. Remember that you'll need long johns as well as upper-body protection.

Outerwear A well-insulated jacket is a must, as are insulated trousers or salopettes in cold conditions. If the weather is likely to be wet you'll need waterproofs; don't take unwaterproofed down out in the rain as it soon becomes soggy and useless. Some local suppliers such as snowmobile operators and the folks at the ICEHOTEL will loan one-piece thermal suits to put on over your jacket and trousers.

Gloves In very cold weather, it's a good idea to wear two pairs of gloves – one thick pair of mitts (mitts that don't separate the fingers keep your hands warmer) and a thin pair of gloves underneath that allow you the use of your fingers when you need

At Sweden's ICEHOTEL even the furniture is sculpted from ice (RTHS/DTW) page 39

to do something fiddly, yet ward off the icy cold for a short time, at least. If you're prone to cold extremities, you can buy carbon hand- and foot-warmers from most good outdoor shops. Shake these up to activate them, pop them into your glove or boot, and they stay warm for around eight hours. If you're going to stay outdoors for an extended period, pack a spare pair of gloves – if you lose one in cold temperatures you'll soon freeze your fingers.

Footwear You'll need proper winter boots if you're going to be outside for extended periods. Many local suppliers will provide these – make sure you request a size larger than you normally wear, to comfortably accommodate extra pairs of socks. Hiking-style winter boots are suitable for simple excursions such as northern lights viewing and town-based activities, but they're not advisable for more adventurous snow-based activities as snow can easily get inside them. Make sure your footwear has good grip for walking on snow and ice.

Socks These should be made of wool, never cotton. Pack an extra pair or two in your rucksack if you're going out snowmobiling, skiing or dog sledding – if your feet become damp or wet you should change into dry socks: wet feet soon become frostbitten feet.

Hats and headwear Take a woollen or fleece hat which covers the ears, as well as a balaclava, Buff or face mask to cover mouth, nose and cheeks. Noses and cheeks are especially prone to frostbite, and should be kept covered whenever possible – skin can freeze in minutes in very cold weather.

Eyewear You may need sunglasses or tinted goggles as the Sun on the snow can be dazzling. Contact-lens wearers may find the cold and dryness makes lens-wearing uncomfortable and should pack glasses as an alternative.

Swimwear If you intend to use any sauna facilities, you may want to pack a swimsuit – unless you're brave enough to go 'au naturel'. There may be occasions too when you'll find yourself in an outdoor hot tub watching the northern lights overhead.

Cosmetics The northern air is very dry, and you'll need to pack plenty of lip salve. Some people have problems with water-based moisturisers. Specialist products are available – ask your travel operator or local chemist.

TEMPERATURES

To give you an idea of the temperatures you'll face whilst in search of the northern lights, the average daily temperatures in degrees Celsius are listed below for Kiruna, Swedish Lapland, home of the ICEHOTEL. Remember, though, that the wind-chill factor can reduce these figures.

Kiruna Dec −13°C / Jan −14°C / Feb −12°C / Mar −9°C / Apr −3°C

FROSTBITE This occurs when the skin and underlying tissue freeze due to extended exposure to very low temperatures. It can affect any part of your body, but the extremities, such as the hands, feet, ears, nose and lips, are the areas that are most likely to be affected. Nevertheless, by wearing the right clothing and taking sensible precautions frostbite can be avoided.

Winter driving holidays are typical in Iceland, but a hardy vehicle such as a Superjeep is crucial for off-road excursions in extreme conditions (S/DTW)

WINTER DRIVING

Although most short winter breaks, especially to Lapland, do not need or include car hire it is possible to hire a suitable vehicle and take off on a self-drive excursion. In Iceland, winter driving breaks are more typical. Wherever you may be, bear the following in mind:

- Take advice from a specialist travel company or local car hire firm when choosing which type of vehicle to hire. They will also have information on which areas of your chosen destination are accessible and which to avoid.
- Take local advice before setting out. If you're driving your own vehicle, make sure it is winterised. Always carry winter clothing and plenty of food as well as an emergency survival kit that includes a torch or headlamp, a shovel, sand or cat litter, and a survival candle. Mobile phone coverage may not stretch to remote areas and sometimes another vehicle will not come by for a long time. If the temperature is –30° or –40°C, if you have an accident and you can't keep your vehicle's engine running, and if you're not carrying proper winter clothing, you'll freeze fast.
- Drive slowly and according to the local road conditions. By law, car rental companies across the region are required to fit winter tyres for improved grip. Some, but not all, will be studded. You'll soon know which you have once you pull away – the studded variety make a faint tick, tick, tick sound on tarmac. Car hire companies also offer a support system in case of breakdowns.
- Keep the tank as full as possible and always refuel when you have the opportunity.
- If you do have an accident or breakdown and need to keep the engine running to keep warm, make sure the exhaust pipe is not blocked with snow and always keep a window open to let in fresh air, otherwise there is the danger of accidental asphyxiation. You may prefer to burn a survival candle than to keep the vehicle's engine running.
- Lastly, if you pass a broken-down vehicle, you *must* stop. Your failure to help could be critical.

HOW TO PHOTOGRAPH THE NORTHERN LIGHTS

With compact cameras, photographing the northern lights is often a matter of luck, although a night-time setting sometimes helps. Capturing that perfect shot can be surprisingly easy, however, if you use an SLR camera that allows long exposures of 10 to 20 seconds together with a tripod. For the best results you'll also want a lens with a wide aperture (f2.8 is good enough, f2.4 is better and f1.4 is best) and a wide angle. Experienced northern lights photographers often wrap foam around their tripod's legs to prevent themselves from touching the metal with bare fingers (the skin sticks). Many photographers prefer to use a cable release – but just pressing the button can also work well. It's also a good idea to wear a pair of thin gloves to protect your skin from frostbite – but remember the protection thin gloves offer will last only a few minutes in cold conditions. Carry thick mitts, too. Plus, you'll need to take a headlamp so you can see what you're doing in the dark.

Camera batteries die fast in cold temperatures. Always carry a spare, and keep it tucked into your clothing, close to your skin to keep it warm. Even better, buy a battery grip (they cost from about £100) and load it with lithium AA batteries – these keep their charge for a reasonable period. Try not to take your camera indoors with you as the lens will fog up with the change in temperature and, when you go outside again, the condensation will turn to ice. If you must take your camera indoors, put it in a plastic bag – ziplock seals work well.

When selecting a location for your photography, try to find a place with some foreground – a tent or building lit from the inside, or some trees – that will give your pictures perspective. For the best outcome, an ISO of 400 is probably best; go up to 800 and the photos can be noisy. Set the focus to infinity and open up the aperture as wide as your lens will allow. If you can, turn off the LCD display as its light will interfere with your vision through the viewfinder in the dark. Then just point and shoot and pray.

One last word: unless you have a fisheye lens, your camera will capture only a fragment of the sky, and during a really spectacular display, the northern lights work their magic across 360°. It's easy to get carried away with photography – but for the best experience, remember to take a few moments just to step back and enjoy the show.

MAKING MOVIES It is virtually impossible to film the northern lights using a regular movie recorder. To record moving images of the northern lights, expensive specialist equipment is required.

NORTHERN LIGHTS FORECASTING

Northern lights forecasting is generally accurate – it's much more reliable than the weather forecast. The forecast corresponds to the planetary magnetic index (Kp) on a scale of one to nine, with one being very low activity and nine a humdinging dazzler. The Geophysical Institute at the University of Alaska has an excellent website (*www.gi.alaska.edu/auroraforecast*), which allows you to view predicted activity in all auroral regions. You can also sign up for email alerts that tell you when activity rises above four to five on the Kp scale.

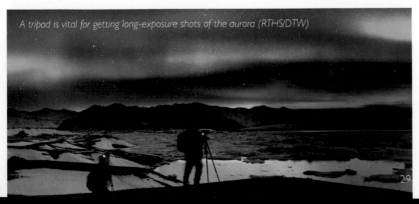
A tripod is vital for getting long-exposure shots of the aurora (RTHS/DTW)

Cross-country skiing is popular on Moskenes Island, at the southern tip of Norway's Lofoten archipelago (KFO/DTW)

At the Santa Claus Village in Finland, you can see Father Christmas and his elves alongside the aurora (VR) page 35

Snowmobiling is a popular daytime activity for aurora-watchers (MC/DTW)

Iceland's Blue Lagoon is conveniently close to the airport, and is a popular attraction for visitors (RTHS/DTW) page 37

4

Beyond the Northern Lights

But what will you do in the daytime? The northern lights are visible only during the hours of darkness, and are best seen in rural locations away from the bright lights of inhabited areas. Some people will while away the day reading an improving book. For the rest, entrepreneurial northerners have created a whole winter wonderland of diversions.

SOMETHING FOR THE DAYTIME

DOG-SLEDDING Dog-sledding trips are a bit like pieces of string: you can cut them as long or as short as you like. Most Scandinavian resorts offer outings of an hour or two, where you'll sit in the sledge and an experienced musher will drive you through snow-laden forests and across frozen lakes, on the lookout for Arctic foxes, ptarmigan or even elk if you're lucky. Some excursions let you buddy up and take it in turns to mush. The more adventurous, however, might like to opt for a longer, hands-on experience. On such specialist tours, you'll mush your own team across various trails and terrain and really get to know your dogs, looking after them over a period of three to four days or more. The longer you can spend with your dogs, the more you'll gain from the experience.

SNOWMOBILING Snowmobiles, or skidoos, are a bit like jet skis on snow. You'll soon get to know the sound they make and once you've opened up the throttle after a little instruction you'll feel like a local in no time. Snowmobiles provide a useful way of travelling efficiently across frozen landscapes and they allow visitors to see much more of their surroundings than they would if they stick to the road network.

Trips can last anything from an hour to a day or even longer, and in many locations special evening snowmobile trips take you in search of the aurora. For those who are afraid of the cold, they are warmer by far than a dog sled (they often have heated handlebars, and the engine helps keep you toasty). You will need a full driving licence to drive a snowmobile.

CROSS-COUNTRY SKIING If you want to combine your northern lights viewing with cross-country skiing in the daytime, two top destinations are Abisko in Sweden (page 13) and Ylläs in Finland (page 15).

Abisko sits at the beginning of the legendary King's Trail, or Kungsleden as it is known in Swedish, which runs for 440km southwards to Hemavan. The trail passes through the pristine wilderness of four national parks and a nature reserve. Most people take a few weeks to ski the whole shebang but you can just do a section – the most popular is the first 86km between Abisko and Kebnekaise. Accommodation is in cabins dotted along the way; various tour operators offer supported trips for those who'd prefer not to go it alone.

There are 330km of maintained cross-country trails in Finland's Ylläs; 38km are illuminated during the dark months. Many of the trails fall within the Pallas-Ylläs National Park. Cafés sit along many of the routes, and lessons are available in both Äkäslompolo and Ylläsjärvi, the villages that serve the area.

Cross-country skiing is a way of life for the locals in Alaska and northern Canada where frozen lakes and rivers, and summer hiking trails, all make great ski routes once the snow falls. All the major destinations covered by this book have cross-country ski clubs that can advise on trails and conditions.

ALPINE SKIING Scandinavian ski resorts tend to be smaller and less developed than those of the Alps – but the great advantages are that the snow is reliable, the pistes are uncrowded, and you can combine your downhill endeavours with northern lights viewing, dog sledding and even a visit to Santa.

Björkliden in Sweden has 23 runs and good off-piste opportunities. It's conveniently close to Abisko National Park from where you can take a chair-lift ride up to the Aurora Sky Station (page 13) for northern lights viewing and the ICEHOTEL (page 39) in Jukkasjärvi. You can either stay in Björkliden and drive over to Abisko in the evenings, or stay at the Abisko Mountain Station and ski down into the Björkliden trail system from the top of Abisko's chair-lift. Also nearby are the downhill resorts of Riksgränsen and Narvik, while the area between Abisko and Narvik offers fantastic ski touring possibilities. Mount Nuolja is recommended for off-piste enthusiasts. Another good option for downhill skiing is the Ylläs area of Finland (page 15), which has 62 slopes and 30 lifts. It's excellent for off-piste, telemark and cross-country skiing, too.

Alaska's most popular resort is Alyeska, near Anchorage, where lifts open late at the weekend, offering the chance to ski beneath the aurora. Alaska is also known for its extensive heli-skiing and cat-skiing opportunities.

SNOWSHOEING Snowshoes have long been worn by residents of the north to allow them to walk on deep snow without sinking. Flat, outsized plates strapped to your

KIRUNA: A TOWN ON THE MOVE

There's something extraordinary about the Swedish town of Kiruna – the nearest town to the ICEHOTEL. Firstly, it's home to the LKAB mine. This, the largest iron-ore mine in the world, has a staggering 400km of paved, two-lane underground roads and every day excavates sufficient iron ore to build 2,500 cars. But the most amazing thing about Kiruna is not its first mine. Nor the fact that they've discovered another iron ore body directly below the town centre. No, the most surprising thing about this chilly spot 200km above the Arctic Circle is that, in order to get at it, they're moving the town.

Over the next 20 years, about a third of the town and its 20,000 inhabitants will be relocated three kilometres to the east. The residents don't seem too fussed; after all, the town only exists because of the mine, which for the last hundred years has been the major employer in these parts. Still, it's a gargantuan task: schools, houses, hotels, office space, leisure facilities, hospitals and the rest must all be created. The new town looks very different to its former incarnation: narrow streets are designed to help preserve the buildings' heat in Kiruna's sub-zero winters, while straight boulevards hope to channel the glacial winds.

The people will take some of their old buildings with them. Kiruna's church, an ornate wooden construction (see page 11 for a photo), was once voted the best-loved building in the whole of Sweden. It will be dismantled, rafter by delicate rafter, and reconstructed in its new location. The original plan was also to move Kiruna's town hall, which is deemed to be an architectural work of art: it cannot be pulled down. The mine's authorities hoped to relocate the town hall in a single move, atop 50 or so outsized lorries. In order for these lorries to progress the few kilometres to the town hall's new location, they'd have had to specially construct an 80m-wide road. In the end though, the task was thought too large. A new town hall has been designed, and just the clock tower will go.

The Aurora Sky Station at Abisko National Park is a top spot for aurora watching due to the large number of cloudless nights (RTHS/DTW) page 13

- The Sámi's traditional lands span the northern parts of Norway, Sweden, Finland and Russia's Kola Peninsula but, even in these areas, the Sámi are now a minority. The Sámi refer to their land as Sápmi.
- The Sámi were once called the Lapps, which means 'scraps of cloth' in Scandinavian languages. Many Sámi people consider the term to be derogatory.
- Sámi language and culture declined in the first half of the 20th century, especially in Norway where the Sámi population is greatest, due to assimilation and cultural suppression.
- In recent decades, Sámi language and culture have been promoted. There are separate Sámi parliaments in Norway, Sweden and Finland (Russia does not recognise the Sámi as an ethnic minority group), and in these countries the Sámi people share responsibility for the management of their territorial lands.
- It's tricky to say exactly how many Sámi people live in Scandinavia and western Russia today. Some estimates put the number at around 100,000 while others say the population is just 30,000 or 40,000. About half live in Norway; Sweden has the next largest population.
- In appearance, Sámi look very similar to other Scandinavian people. They are often blond-haired and blue-eyed.
- There are many different Sámi languages. Although the languages are linked – they're related to Finnish and Estonian – not all Sámi can understand each other. Not all Sámi people can speak a Sámi language, but all can speak the language of the country where they live. A few schools across northern Scandinavia teach most classes in the Sámi language.
- Traditional Sámi industries include reindeer husbandry, hunting and fishing – but only 10% of Sámi herd reindeer today. Modern-day reindeer herding uses snowmobiles, helicopters, mobile phones and lorries to transport the stock.
- The Sámi flag is vibrantly coloured, with a half-blue, half-red circle over a background of red, yellow, green and blue panels and stripes. It was raised for the first time in August 1986.
- Traditional Sámi dress varies between regions, but often consists of bright blue wool or felt tunics with red and yellow embroidery and bands for men, and dresses for women in a similar style. Both men and women wear reindeer-skin boots.

Sámi language and culture have been promoted in Norway, Sweden and Finland in recent years, and visitors to the far north can visit their traditional conical tents (RTHS/DTW)

Mývatn Nature Baths are one of the most popular hot springs in Iceland (MNB)

boots, they spread your weight over a wider area, keeping you above rather than in the snow. The snowshoes of bygone days looked like giant tennis rackets, but now they come in all colours and materials. A tourist snowshoeing excursion will generally involve going for a hike with a guide who will point out animal tracks, local points of interest and share stories of the area.

HORSERIDING The Icelanders are proud of their horses, and determined to keep the breeding of Icelandic horses pure: no horses can be imported into the country and an Icelandic horse, once exported, may not return. Visually distinctive, the horses possess two unique gaits as a natural addition to the standard walk, trot and canter. Visitors to Iceland can take horseriding excursions throughout the year. During the winter months, they'll be provided with insulated clothing. Visitors to the ICEHOTEL in Sweden can also ride an Icelandic horse to see elk during the day and hopefully the aurora in the evening. These horses are suitable for novices and children over the age of 12.

VOLCANOES The mid-Atlantic ridge cuts neatly through the middle of Iceland. This is where the Eurasian and North American tectonic plates are pulling apart at a rate of just over 2cm a year, giving rise to the geothermal liveliness for which the country is famous. Sightseeing flights over the volcanoes, including the famously disruptive and unpronounceable Eyjafjallajökull, take to the skies regularly, weather permitting.

VISITING SANTA CLAUS The Americans may claim that Santa comes from the town of North Pole, Alaska (named in the 1950s by a development company that hoped to attract a toy factory to its rather chilly real estate), and the Russians may insist that he's theirs but, deep down, everyone knows that Santa lives in Lapland. Of all the Scandinavian nations, it's the Finns who trade most enthusiastically on the Christmas connection. Following a myth engendered by the Finnish Broadcasting Corporation in the early decades of the 20th century, Finns will tell you that Father Christmas comes from Korvatunturi, a mountain in Urho Kekkonen National Park. But Korvatunturi is impractical for money-making folk – it's just too remote – so the city of Rovaniemi cashed in on its relative geographical proximity and declared itself the 'official homeland of Santa Claus'. Now there's a Santa Claus Village, including

My reindeer was called Girjebahta, which means 'Spotty Bottom' in the Sámi language. She was a beautiful creature with pale fur, long, elegantly arching antlers – and an attitude that, had Santa been foolish enough to select her, would have put the kybosh on Christmas.

I was with a group of fellow reindeer novitiates in Swedish Lapland. Our Sámi guide, Nils-Erik, had shown us how to lasso the reindeer's antlers with bright orange rubber rope. Then he'd harnessed one creature to each sled, instructed us to kneel one person to each, and to make a circuit of the track that ran between the snow-laden spruce trees. But I'd only covered a few metres before Spotty ran out of steam and, rather than running, pretended to nibble at a sparse little twig that lay on the path before her.

'Woop woop,' I shouted (*woop woop* is Sámi for 'get a move on') and I whacked her on the bottom with the rope. Spotty turned and pouted, and planted her pretty little hooves in the snow. But then a sprightlier reindeer appeared from behind and Spotty's competitive spirit was spurred.

Neck and neck, the two animals galloped. Our wooden sleds leapt from the peaks of ruts in the snow, and clattered into its divots. Antlers almost clashed, sleds seemed certain to derail, and as the opposing team tried to squeeze by on the inside I had to duck low to avoid a disembowelment. Then, just when disaster seemed inevitable, Spotty stopped stock still and refused to move once more.

'You have to show the reindeer who's boss,' Nils-Erik explained wearily, having trudged out around the track to lead us in. I quietly suspected that both my reindeer and I knew which of us that was.

Later, we sat in a *lavvu* – the large, conical Sámi tent – and ate reindeer meat that Nils-Erik had fried in a gargantuan pan over a fire, followed by stewed loganberries. There are around 200,000 reindeer in Sweden, Nils-Erik told us, and almost all of them are domesticated by the country's 20,000-strong Sámi population. Their antlers grow in summer – as much as a centimetre each day – and then drop off in autumn and winter.

Nils-Erik had spent the previous four days herding reindeer. He and his companions had separated the animals into the smaller groups that belong to each family – his parents owned a thousand or so of them. It's hard physical work; Nils-Erik revealed that he went to the gym to toughen up for the reindeer-herding season. But the Sámi people nowadays use snowmobiles to travel and, where feasible, they even truck their reindeer by road. As for the *lavvu*, nobody sleeps in one any more – which was just as well for Nils-Erik.

'I'm allergic to the reindeer skins that cover the ground,' he told us, 'and the smoke from the fire sets off my asthma.'

Reindeer-sledding trips are a great way to explore the open countryside (VR)

Santa's main post office, a 'Santapark' theme park, a bunch of elves and a sled-load of reindeer. And if that's not enough to draw you to Rovaniemi, get this: it's also home to the world's northernmost branch of McDonald's.

SAUNAS, SPAS AND HOT SPRINGS The Finns just can't help themselves – there's nothing they like more, it seems, than to take off their clothes, get very sweaty, and then go for a roll in the snow. The good news is, tourists can indulge in the experience, too (for many, it's a highlight of their trip). Almost all accommodation comes complete with a **sauna** and some have their own private hot tub so your fellow guests needn't see your flabby bits. In Alaska, **Chena Hot Springs** are open daily till midnight, while **Takhini Hot Springs** in Whitehorse rents its pools to private groups from 22.00 till 01.00 – and for those concerned with their coiffure, there's a hair-freezing contest. Sculpted by volcanic eruptions across the ages, the Mývatn area of Iceland has a striking beauty defined by craters and lava columns, sulphurous steam vents and hot pools that bubble and burp. Visitors can bathe in the **Mývatn Nature Baths**, whose milky-blue lagoon keeps a year-round temperature of 38°–40°C. The **Blue Lagoon**, with its steamy pastel-blue pool set amid a jet-black lavascape, is one of Iceland's most iconic tourist attractions. Water temperatures remain at 37°–39°C year-round, and spa treatments and massages are available. The nearby Northern Light Inn is within walking distance, and its 360° observation lounge is a great place from which to watch the aurora doing its thing above the lavascape.

WINTER WEDDINGS White stilettos might be out of the question, and brides: don't forget your faux-fur stole. But dress properly for the occasion and an ice chapel venue can take your wedding to a whole new shade of dazzling white. The ICEHOTEL in Sweden (page 39) builds its own ice chapel each winter.

Glacier walking is one of the activities available from Hotel Rangá in Iceland (HR/DTW) pages 39–40

POLAR BEARS Churchill, Manitoba is the polar bear capital of the world: nearly 1,000 bears gather here in October and November as they wait for Hudson Bay to freeze so that they can hunt the seals that live beneath its ice. Visitors can travel out to see polar bears in specially designed 'tundra buggies', or even take a trip to the polar bear jail whose inmates have strayed too far into town.

DOG-SLED RACES Two of the mushing world's major races take place in Alaska and Yukon during the winter months. In February, the thousand-mile Yukon Quest runs between Whitehorse and Fairbanks, while in March the Iditarod covers the same distance between Anchorage and Nome. The Iditarod celebrates the serum run of 1925: when diphtheria broke out in the coastal village of Nome, doctors feared an epidemic. The only way to save the village was to transport serum by dog-sled relay, day and night over 700 miles of frozen wilderness. (Gay and Laney Salisbury's telling of the story in their brilliant book *The Cruellest Miles* is a must-read for anyone interested in Alaska and its history; page 47). Visitors can spectate from the start and finish, and various points along the routes of both races including the wonderfully comfortable Winterlake Lodge (page 40). Go to www.yukonquest.com and www.iditarod.com for more information.

GOLD-RUSH HISTORY Alaska and northern Canada are rich with gold-rush history – many of the towns owe their very existence to the gold rushes and the infrastructure that sprung up to house and feed the stampeders. The big one was the Klondike, in Canada's Yukon, to which an estimated 100,000 hopefuls swarmed in 1898. Visitors to Whitehorse can find out more at the MacBride Museum or, better still, add a few days to their trip by taking in Dawson City, the town that was created by the Klondike Gold Rush and whose historic charm remains almost unspoilt.

The Yukon Quest is a 1,000-mile dog-sled race that runs between Whitehorse and Fairbanks each February (TY)

AND IF YOU HAVE TIME TO SLEEP

From ice hotels to igloos to secluded log cabins with roaring wood stoves, when in search of the lights you'll have a variety of accommodation options – although you might have to clomp to the bathroom in big winter boots rather than mince elegantly in complimentary towelling slippers. Here are a few of the more interesting options.

ICEHOTEL, SWEDEN When Yngve Bergqvist first came up with the idea of building a hotel entirely from ice in northern Sweden, everyone said he was barmy. Who on earth would want to sleep in a hotel made from ice, his incredulous friends and associates asked. The bank managers, meanwhile, stared glacially from behind their orderly Scandinavian spreadsheets and uttered a stark 'Nej'.

There was little funding. And so, in the late 1980s, the ICEHOTEL started small. Over the years, however, it's snowballed: there are now franchised Icebars from Tokyo to Shanghai, Stockholm, Copenhagen and London. At Jukkasjärvi, where the ICEHOTEL was born, there are now two ICEHOTELS: ICEHOTEL 365, which is open year-round, and the winter-only ICEHOTEL whose number changes each year. In winter 2017–18 it will be ICEHOTEL 28; the following year will see the creation of ICEHOTEL 29. There's also a whole village of warm facilities – Kaamos rooms, two types of chalet, two restaurants, a lounge/bar and champagne bar – in which the chilly visitor can find respite.

But it's still the sub-zero section of the ICEHOTEL that takes the breath away. The ice from the Torne River, from which the hotels are built – and back into which the winter ICEHOTEL melts gracefully each spring – is known for its purity and clarity. It's said to be better to drink than bottled mineral water. Maybe it's this, together with clever use of lighting, that makes this magnificent place more art gallery than hotel. Around 30,000 litres of water is used in the construction of the hotels – that's 700 million snowballs to those whose minds can stretch to such a thing. The chandeliers alone comprise more than 1,000 hand-polished ice crystals. In the Icebar, even the glasses are made from ice – they'll slip from your hands if you're not wearing gloves. But unlike an ice cube, they're kept at a temperature just warm enough that they don't stick to your lips. Rather, the glass gently melts as you drink, moulding itself to the shape of your mouth.

In the winter ICEHOTEL, temperatures hover around –5° C in the simple Ice Rooms, where you sleep in thick sleeping bags on reindeer skins and a mattress on top of a wooden slatted base which slots into the ice-carved bed frame. These rooms are almost monastic in their simplicity, and utterly silent between their thick snow walls. But step across the hallway, through the avenue of glistening sculptures, and you'll come to the Art Suites. Here, individual artists from countries across the world have each designed and created a bedroom. They're different every year. When I visited, there was a thrusting tango suite from Argentina, red-hot and sultry despite the ice, a delicate Japanese garden whose tiny details blended with the muted white tones, and Maori *moko* carved out of the ice by their Kiwi creator.

In ICEHOTEL 365, you choose between an Art Suite or one of nine Deluxe Suites, all of which have a bathroom. Some also come equipped with sauna. And in summer? The ICEHOTEL harnesses the energy of the midnight sun. Power generated by solar panels keeps the hotel cool while its roof is insulated by grass and Arctic flowers.

HOTEL RANGÁ, ICELAND Hotel Rangá's rural situation means that no light pollution interferes with your aurora viewing; its geographical location means that it enjoys vast open skies with 360° views, with mountains, glaciers and volcanoes providing a distant backdrop. Guests can lie back and watch the aurora from the hotel's naturally heated outdoor hot tubs and, if requested, staff will wake guests during the night when the lights appear – so there's no need to wait rubbing your eyes into the small hours. For volcano enthusiasts, the hotel sits in the so-called Ring of Fire, where geothermal activity merrily spurts and bubbles. It may be possible to take a sightseeing flight from here, weather permitting, as well as 4x4 excursions exploring the many places of interest in the area. The hotel is close to **Skogafoss and Seljalandsfoss**, two of Iceland's iconic waterfalls. The Golden Circle area is within easy reach, where visitors can tour spouting geysers, volcanic rifts and another waterfall, two-tiered **Gullfoss**, which sometimes freezes spectacularly in winter. For a longer day trip, **Jökulsárlón glacial lagoon** is filled with icebergs jostling for position as they head out to sea.

HOTEL HÚSAFELL Just a 90-minute drive north of Reykjavik, the contemporary Hotel Húsafell lies at the foot of the magnificent Borgarfjörður mountain range, whose peaks help to create those cloudless nights so cherished by aurora seekers. They see the northern lights about three times a week here during the winter months. And when you're not standing with your head turned towards the heavens, there's a fabulous outdoor spa that includes float water therapy, as well as a renowned restaurant specialising in local fare. Daytime excursions include a drive by snow truck onto the Langjökull icecap, from where you walk on foot into the heart of the glacier through a brilliant-blue ice tunnel.

HOTEL SIGLÓ This is a brand-new hotel in a historic spot: it sits right on the waterfront in Siglufjörður, once a tiny shark-fishing village, then a herring hotspot. Now the herring have gone, but the town still relies on the daily catch, which you can watch being brought in – and you can eat at any of the three restaurants in the Marina Village. The northern lights have a particularly scenic backdrop here, against the village's mountain and ocean setting, and the hotel's hot tubs will keep your toes toasty as you stare into the skies.

REYKJAVÍK, ICELAND The nation's capital is a great place for a winter getaway, with the Blue Lagoon and plenty of other attractions on its doorstep, some of them reached by Superjeep. And surprisingly, even with the glow from the city, northern lights sightings out across the bay can be amazing.

WINTERLAKE LODGE, ALASKA Winterlake Lodge is one of the world's magical spots. Truly remote, yet properly luxurious, it's only accessible by ski plane in winter (and float plane in summer). Owners Carl and Kirsten Dixon keep sled dogs and offer mushing trips that take in the famous Iditarod trail, on which the lodge sits (page 38). It's a checkpoint for the Iditarod race, as well as other dog-sled competitions, so the ideal location from which to soak in some of the atmosphere of the race. Cross-country skiing, snowshoeing and winter camping are also available, while indoors there are daily cooking classes, yoga and massage – not to mention Kirsten's legendary cooking. Come nightfall, the hot tub on the terrace makes an ideal spot for lying back and watching the northern lights.

BLACHFORD LAKE LODGE, CANADA Like Winterlake in Alaska, Blachford Lodge, near Yellowknife in Canada's Northwest Territories, is only accessible by ski or float plane. Daytime activities include snowmobiling, skiing, snowshoeing, ice fishing, skating or relaxing in the hot tub. Again, you can watch the aurora from the hot tub – solar flares permitting. Prince William and Kate visited here in 2011, and if it's good enough for them…

5

The Northern Lights in History

EXPLORERS AND THE AURORA

In 1815, the Napoleonic Wars came to an end with the Battle of Waterloo. For the last few decades, officers of the British Navy had been gainfully employed trying to blast the French and the Americans off the face of the ocean; now that hostilities had ended, the Navy found itself with a surplus of officers who had to be retained on half pay. **John Barrow**, Second Secretary to the Admiralty, was blessed with a yearning for exploration – as well as a burning desire to do something useful with his surplus of men. He therefore rolled out his maps and set himself the task of filling in the blanks.

So began a great era of discovery. Barrow sent expeditions into the heart of Africa and to Antarctica (for more information read the excellent *Barrow's Boys* by Fergus Fleming – page 47), and he dispatched officers to the northern extremes of the globe: two of his missions were to find a trade route through the Northwest Passage, and to determine what lay at the North Pole. The naval officers heading these voyages were educated men. As they travelled they mapped coastlines, took scientific measurements and recorded their observations. Their experiences and theories relating to the northern lights added greatly to contemporary understanding of the phenomenon.

John Franklin – famous mainly for disappearing into the icy white, although his mapping of the North American coastline was a significant achievement – confirmed that appearances of the aurora were connected to magnetic activity. 'My opinion recorded in my former Narrative, that the different positions of the Aurora have a considerable influence upon the direction of the Magnetic Needle, has been repeatedly confirmed during our residence at Bear Lake. It was also remarked, that, from whatever point the flow of light, or, in other words, the motion of the Aurora proceeded, if that motion was rapid, the nearest end of the needle was drawn towards that point almost simultaneously with the commencement of the motion', Franklin wrote in his *Narrative of a Second Expedition to the Shores of the Polar Sea in the Years 1825, 1826 and 1827*.

After Barrow and his fur-clad missionaries passed on, a new generation of explorers followed in their wake, still puzzling over the same geographic conundrums. The Norwegian scientist and explorer **Fridtjof Nansen** was the first man to cross Greenland by ski; then in 1893 he sailed to the Arctic aboard the *Fram* and reached the highest latitude of any explorer yet, at 86°14'N. Nansen wrote lyrically about his observations of the northern lights. 'However often we see this weird play of light, we never tire of gazing at it', he wrote in his journal on 22 December 1895; he later published his account in a book titled *Farthest North*.

It seems to cast a spell over both sight and sense till it is impossible to tear oneself away. It begins to dawn with a pale, yellow, spectral light behind the mountain to the east, like a reflection of a fire far away. It broadens, and soon the whole of the eastern sky is one

glowing mass of fire. Now it fades again… After a while, scattered rays suddenly shoot
up from the fiery mist, almost reaching to the zenith; then more; they play over the
belt in a wild chase from east to west. They seem to be always darting nearer from a
long, long way off. But suddenly a perfect veil of rays showers from the zenith out over
the northern sky; they are so fine and bright, like the finest of glittering silver threads.

In 1905, Nansen's compatriot **Roald Amundsen** was the first man to succeed in
negotiating the Northwest Passage; six years later he was also the first to reach the
South Pole. He, too, recorded his observations of the aurora – and he was convinced
that he heard them, as he wrote in his book *The South Pole*.

The light is so wonderful; what causes this strange glow? … It is one of the few really
intense appearances of the aurora australis that receives us now. It looks as though
Nature wished to honour our guests, and to show herself in her best attire. And it is a
gorgeous dress she has chosen. Perfectly calm, clear with a starry sparkle, and not a
sound in any direction. But wait: what is that? Like a stream of fire the light shoots
across the sky, and a whistling sound follows the movement. Hush! Can't you hear?
It shoots forward again, takes the form of a band, and glows in rays of red and green.
It stands still for a moment, thinking of what direction it shall take, and then away
again, followed by an intermittent whistling sound. So Nature has offered us on this
wonderful morning one of her most mysterious, most incomprehensible, phenomena
– the audible southern light.

A bust of Roald Amundsen on Svalbard, Norway. He was enthralled by the aurora at both the North and South poles (AS/WC) page 8

KRISTIAN BIRKELAND

Kristian Birkeland was the first man to claim that the aurora's source is the Sun. He correctly insisted that the Earth is surrounded by magnetic fields, which guide solar particles to the polar regions, where they produce light in the form of the aurora in the Earth's high atmosphere.

A bald, bespectacled man physically weakened by overwork and insomnia, Birkeland was born in Christiania (now Oslo) in Norway in 1867. Obsessed with auroral observation, he spent the winter of 1899–1900 cooped up in a specially built observatory on Haldde Mountain in the far north of Norway. When he wasn't being buffeted by frigid storms, he took ceaseless magnetic recordings of the aurora, and concluded that it is caused by electric currents in the Earth's high atmosphere that emanate from the Sun and are directed towards the polar regions; that these currents run at least 100km above the Earth's surface; and that magnetic disturbance is a result of these electric currents rather than a direct cause of the northern lights.

When Birkeland's leadership of a second research expedition to Russia was delayed by a dog bite and the need to recuperate from the side effects of a rabies vaccination, Birkeland used his enforced period of rest to think further about the aurora. He became convinced that, to fully understand it, he must recreate the aurora in laboratory conditions; this way he could be free not only from the risks posed by bad-tempered dogs but from the inconveniences of the Arctic weather.

An inveterate aurora enthusiast, Kristian Birkeland (1867–1917) was the first to realise that the northern lights are caused by the Sun. Today he and his inventions are honoured on the Norwegian 200-kroner banknote (SS) page 43

His problem was money: the laboratory would cost more than he could possibly raise through his normal channels of funding. And so he tried his brain at business. He invented an electromagnetic cannon that, he claimed, would be able to fire from a ship further and faster than any weapon before. Unfortunately, the much-vaunted public unveiling of Birkeland's cannon was a disaster. His current short-circuited with a burst of flame and a fearsome roar, and shot an arc of brilliant white light across the auditorium from which some of the terrified audience fled. It also left a strange smell in the air, a bit like that which lingers after a lightning flash.

Plans for the cannon were off, but its dramatic failure gave Birkeland a new idea. The smell – created both by the faulty cannon and by lightning – was that of nitrogen oxidising. If Birkeland could reproduce the reaction, it would effectively mean he could manufacture lightning – and the nitrogen–oxygen blend it produced would allow him to create fertiliser (a nitrogen–oxygen combination) on an industrial scale.

For three years Birkeland spent every waking minute working on a furnace to produce fertiliser. Then he sold his business interests and was free to work on his auroral theory. He built a vacuum tube, which he called a 'terrella'. Inside was a ball – standing in for the Earth – whose magnetism he could turn on and off. With conditions inside the tube created to mimic the Earth in space, he flooded the tube with electrons from a cathode. When he switched on the magnet, the electrons ran towards the 'Earth's' polar areas, and created rings – like the auroral ovals – around its magnetic north and south.

With his experiment, Birkeland demonstrated his idea that charged solar particles are directed by the Earth's magnetic field, which is comet-shaped with a tail tapering from the night-time side of the Earth, and that it's this that creates the northern lights. His work was so significant to astronomy that even today a portrait of Birkeland and a depiction of his terrella illustrate Norway's 200-kroner banknote, and the term 'Birkeland current' is still used to describe this particular field-aligned current.

But, just as it had made him, Birkeland's genius finished him. His marriage to the long-suffering Ida lasted six years; he was frequently absent due to work and, consumed by loneliness, she finally left him. An inveterate insomniac, Birkeland became addicted to the barbiturate veronal, which induced paranoia: towards the end of his life he became convinced that British spies were trying to steal his cannon invention, and ultimately that they wanted to kill him. He dismissed his domestic staff in case they might connive. He drank heavily. He travelled to Egypt and then to Japan where he worked hard, then feverishly, locking himself in his hotel room for weeks. On the morning of 15 June 1917 hotel staff found Birkeland dead. A postmortem concluded that he had taken 20 times the recommended dose of veronal. He was just 49.

GREAT AURORAS IN HISTORY

12 SEPTEMBER 1621 This was the show that gave the northern lights their name. Exactly which scientist – Galileo or Gassendi – was responsible for the term 'aurora borealis' may be open to debate (page 6). The intensity of the event that inspired them is not: this aurora was powerful enough to be seen as far south as France and Italy.

6 MARCH 1716 The great aurora of spring 1716 was observed by British astronomer Sir Edmund Halley; his decades-long desire to see a display of the aurora borealis was not rewarded until he was 60 years old. Halley was asked by the Royal Society to write up his observations in its publication *Philosophical Transactions*. This was the first detailed description of the aurora. Halley wrote, 'Auroral rays are due to the particles which are affected by the magnetic field; the rays are parallel to Earth's magnetic field.' Halley also correctly theorised that the corona form of the aurora

MYTHS AND LEGENDS

Before the likes of Halley and Birkeland came along with their magnetic fields and their solar flares, the indigenous people of the north attributed all manner of powers and causes to the northern lights.

- Stories of warfare are often connected to the northern lights. Norse legend says that the lights are the glinting shields of Valkyrie warriors carrying their slain enemies to Valhalla, the hall of the dead, while intense displays of the aurora in southern Europe – where they are seen as blood-red – have been considered to be ominous portents of war.
- The aurora is widely connected to the game of football in Inuit legend. Some stories relate that the lights are a manifestation of human spirits playing football with a walrus head; others say that walrus spirits are playing football with a human skull. Some even reckon that they're playing football with the heads of naughty children. Certain Inuit people used to rush indoors when they saw the lights, terrified the aurora would slice off their heads to use as a new ball. Others carried knives to protect themselves from the wrathful spirits.
- Many legends attribute the northern lights to the spirits of the dead. In Iceland, it was believed that the spirits were trying to contact their living relatives, while in northern Canada they used to say that the lights were the spirits dancing to while away the long dark winter, and the colours were their festive clothing.
- It used to be commonly believed that you could bring on the aurora by whistling to it, and that clapping would stop the show. Some northerners thought that whistling or singing to the aurora was dangerous, however, as it enraged the spirits who might then sweep down to Earth and exact their revenge.
- The English legend of St George and the dragon is thought to have originated with the aurora, and many people believe that Chinese dragon legends, too, were born following ancient displays of the northern lights.

was an effect of perspective (page 9). He didn't hit on the source of the aurora, however – he surmised that luminous matter inside the Earth was escaping through cracks in its surface.

28 AUGUST AND 2 SEPTEMBER 1859 Tremendous solar flares shot out from the Sun in August and September 1859. They created the greatest space storms in recorded history. The plasma contained particularly intense magnetic fields and travelled at exceptionally high speeds – it's thought that the second eruption took just over 17 hours to reach the Earth, compared with the usual two to three days. The resulting aurora was seen as far south as Hawaii. Electrical surges disrupted telegraph lines across the northern United States and Canada; the current was so strong that the line between Portland and Boston ran for two hours without batteries, using the power of the aurora alone.

25 JANUARY 1938 'A remarkable and very beautiful appearance of the Aurora Borealis, or Northern Lights, was seen last night from many parts of England, including the South, where the spectacle is seldom to be seen', *The Times* reported from London. 'A Deal fisherman who returned to port last night said: "It appeared as if the whole heavens were on fire, and great beams of red light like steps stretched across the sky."' In Austria, some villagers called out the fire brigade to put out the supposed fires, while in Switzerland the Basel fire brigade was kept on stand-by as it was assumed the light came from fires in neighbouring Alsatian villages. Another *Times* journalist wrote, 'At Zeebrugge the sky was lighted up as if by a huge Bengal flare, and columns of light were seen rising from the sea as if from powerful projectors. The colour changed from red with white columns to blue, and a few seconds after the light had disappeared a large arc was seen over the town.'

13–14 MARCH 1989 The most recent of the great auroral displays delivered such intense currents that in Québec, Canada, the power grid tripped, leaving six million people without electricity for nine hours. Magnificent displays were also observed across Europe.

Appendix: Further Information

BOOKS

Bone, Neil *The Aurora: Sun–Earth Interactions* John Wiley & Sons, 1996. Bone delves deep into the mysterious world of physics, but his book is nonetheless comprehensible to non-scientists. It's detailed and useful for those who want to take their understanding to a higher level.

David, Neil *The Aurora Watcher's Handbook* Chicago University Press, 1992. This has a North American slant and was written back in the 1990s, but even for European readers it's probably still the best guide for the layman wanting a detailed but accessible explanation. If you can't find it on Amazon, order it from *www.uaf.edu/uapress*.

Falck-Ytter, Harald *Aurora: The Northern Lights in Mythology, History and Science* Floris Books, 1999. Falck-Ytter's book is a straightforward read. As the title suggests, it covers legend and history as well as science.

Fleming, Fergus *Barrow's Boys* Granta Books, 2001. This doesn't really have anything to do with the northern lights, but for anyone interested in the exploration of northern parts, including the races for the Northwest Passage and the North Pole, this book makes riveting reading.

Jago, Lucy *The Northern Lights: How One Man Sacrificed Love, Happiness and Sanity to Solve the Mystery of the Aurora Borealis* Penguin Books, 2002. A meticulously researched and highly readable biography of Kristian Birkeland, the Norwegian scientist whose life-long obsession with the aurora resulted in a greatly increased understanding of both the northern lights and of space.

Salisbury, Gay and Laney *The Cruellest Miles* Bloomsbury, 2003. Again, nothing to do with the aurora, but this is a fabulous account of the 1925 serum run, in memory of which the Iditarod race is run.

TELEVISION

In recent years, the BBC has made two excellent, if very different, documentaries relating to the solar system and the northern lights. They are regularly rebroadcast: keep your eye on the TV pages and iPlayer to catch them if you can. The first is the opening episode of the ***Wonders of the Solar System*** series, presented by Brian Cox. The episode is titled '**Empire of the Sun**'. The second is Joanna Lumley's personal account of her own quest for the lights, ***In the Land of the Northern Lights***.

WEBSITES

SOHO (Solar and Heliospheric Observatory) is a space-borne observatory operated by the ESA (European Space Agency) in collaboration with NASA. Even if you don't harbour a deep fascination for the inner workings of the Sun, SOHO's website (*http://sohowww.nascom.nasa.gov*) is worth a visit for its incredible images.

The **University of Alaska Fairbanks Geophysical Institute** has an excellent website. Go to its aurora forecast page (*www.gi.alaska.edu/AuroraForecast*), which includes the forecast for all auroral regions. Click on the links for further information on the northern lights – the FAQ page is particularly helpful. You can also sign up for email alerts from the Geophysical Institute that will tell you when auroral activity in your area is predicted to be high.

Another great website for updates on auroral displays is www.spaceweather.com. Alongside northern lights updates and links to NASA there's also an area where you can submit your own photos of the display. Last but not least, the **British Astronomical Association** (Britain's leading association for the amateur astronomer) has an auroral section on its website with useful tips for observing and photographing the lights, and well-laid-out info on the science behind the aurora (*www.britastro.org/aurora*).